Rationing Beef Cattle

Second Edition

A Practical Manual
by
Dr David Allen
Beef Industry Consultant

Chalcombe Publications

First published in Great Britain by
Chalcombe Publications
Painshall, Church Lane, Welton,
Lincoln, LN2 3LT.
United Kingdom

First Edition 1992
Second Edition 2001

ISBN 0 948617 44 6

Printed in Great Britain by Ruddocks Colour Printers, Lincoln.

Contents

Preface to the Second Edition

When the first edition of this book was published in 1992, I wanted to share with readers the total approach I had developed to feeding beef cattle. It involved setting performance targets, with marketing objectives in mind, for finishing cattle, formulating rations and drawing up feed budgets to ensure that the feeding plan could be carried out with the available feeds. I proposed a novel method of formulating rations that could be carried out with a calculator, pencil and paper without the need for a computer. That implies no criticism of several excellent computer rationing packages that are available.

I never imagined that nine years later there would still be sufficient interest to publish a second edition of the book. Of course, things have moved on. We now have better information on the performance of cattle on a range of diets. Contributing to that is a much better understanding of factors involved in the control of grass silage intake.

The various reports of AFRC Technical Committees were consolidated into an advisory manual, AFRC (1993) 'Energy and protein requirements of ruminants' published by CAB International. The AFRC manual was masterminded by Geoff Alderman who died in 2000 at the age of 75, still busily evaluating energy and protein rationing systems. Without his support and encouragement, I could not have compiled this practical manual.

Since the First Edition there have been major changes in the British beef industry in the wake of the BSE crisis in 1996. When the First Edition was published, the sale price of finished steers averaged £1.11 per kg liveweight and by 1995 it had increased to £1.24. The average in 2000 hovered around £0.85 to 0.90 per kg liveweight. The fall in sale price has been offset by a reduction in input prices. Calves and store cattle are cheaper and the price of feed quality cereals has fallen from around £90 per tonne in 1992 to £60 now. Nevertheless, controlling costs is ever more important, as is managing cattle to perform to the targets set for them. This is why rationing is so important.

The whole purpose of this manual is to encourage forward planning and to offer a simple method by which beef producers and their advisors can set performance targets, design rations and draw up feed budgets. I hope it will help beef producers to squeeze extra hard-earned profit from their cattle.

Dr David Allen
North Coach House
Burcott
Wing
Leighton Buzzard
Bedfordshire
LU7 0LZ

Email:david.allen2@which.net

January 2001

Introduction

Rationing cattle to achieve target levels of daily gain is an essential part of profitable beef production. Devising a ration has four steps:

- Deciding what level of performance is required.
- Formulating the ration.
- Drawing up a feed budget for the whole feeding period.
- Evaluating rationing tactics.

Inevitably, rationing is concerned largely with housed cattle. This is not to imply that the feed requirements of grazing cattle do not matter. But grazing performance depends on managing the relationship between cattle and the sward rather than nutrient allowances *per se*. Nonetheless, drawing up a grassland budget for grazing cattle is just as important as drawing up a feed budget for housed cattle.

An important feature of this manual is that rations can be formulated using a calculator, pencil and paper, which is against the trend to computerised rationing. Of course, the rationing procedures used here can be computerised quite simply.
A shortcoming of many computer-based rationing services is that they produce a ration 'out of the blue' and the four-step approach advocated here is not followed. I believe that it will pay farmers to spend time working out an overall rationing strategy before their advisor visits the farm. Then expensive consultant time can be used to develop the strategy and is not frittered away in routine calculations. If necessary, the computer can be used to check the rations.

The rationing procedure has been kept as simple as possible and minimises the number of look-up tables that have to be used. Taking this simple approach has meant stripping to bare essentials the metabolisable energy (ME) system of rationing (AFRC, 1993).

However, for growing cattle I have preferred a different categorisation of breed and sex effects on ME requirements from AFRC and have therefore presented a full description of assumptions and the calculation of ME allowances in Appendix 1 for perusal by nutrition specialists.

Rationing beef suckler cows aims to manage body condition through the annual production cycle to achieve high reproductive efficiency and rapid calf gains at minimum cost. Whether cows calve in the spring or autumn, in a temperate climate such as that of the UK winter feeding allows some loss of body condition and weight which is restored when cows graze through the summer. The tables of ME allowances in Chapter 3 are for suckler cows in optimum body condition in the autumn but adjustments are proposed for cows that are thinner than intended. The assumptions and method of calculating ME allowances are presented in Appendix 2.

AFRC (1993) proposed a metabolisable protein (MP) rationing system to improve protein rationing. Having studied the new system, I decided to revert to expressing requirements as the crude protein (CP) content needed in the ration and have maintained that stance in this Second Edition. This is not the backward step that it may appear to be at first sight provided the significance of protein degradability is understood and taken into account in the selection of feed ingredients, especially for young cattle up to six months old. Moreover, the diet CP levels recommended here are based on requirements calculated using the MP system that takes account of the energy concentration of the feed.

A brief review of recommended allowances for major vitamins and minerals is presented in Chapter 5.

There is a tendency to become so immersed in the elegant statistical methodology of calculating nutrient requirements and formulating rations that the crucial importance of feed intake is forgotten. One of the main reasons that cattle perform below target is that they are unable to eat the ration.

It has become apparent that no single generalised formula can ever predict intake over the whole range of feeds. The approach adopted here is to predict potential intakes of forages when fed alone, with estimates of the depression of forage intake when concentrates are fed (concentrate substitution). The intake values are generally lower for forages than those presented in the First Edition.

When all is said and done, rationing is an approximate business and formulating the ration is only a first step. Next it is essential that stockworkers know what the ration is and to feed it. Then it is vital to manage the cattle and the feed to ensure that the ration is consumed and the required performance is achieved.

Periodic weighing of growing cattle monitors progress and provides the information needed to revise rations if necessary. No professional beef producer can afford to be without an accurate weighbridge built into an efficient handling system. It is not as necessary to weigh suckler cows but they need to be handled periodically to check body condition.

Chapter 1

Cattle Performance

Narrow profit margins and increasingly demanding market specifications make it imperative to set performance targets for the various categories of cattle. Only then can rations be designed that allow production, marketing and financial targets to be met.

This chapter presents performance standards for growing and finishing cattle, replacement heifers and beef suckler cows.

FINISHING CATTLE

The objectives in setting performance targets for finishing cattle are to decide on the slaughter date and weight, the consequent level of daily gain required and the duration of the feeding period. These pieces of information are a pre-requisite of rationing and feed budgeting.

The best information on which to base performance targets is previous experience with the same category of cattle on the farm. Look through sale returns to find out what liveweights and carcase weights at sale were achieved and whether the EU fat classes * were within the buyer's specification. If possible, from a known start weight calculate the daily gain achieved and the average duration of the feeding period. This information can then be used to help set targets for the present batch of cattle.

In the absence of previous experience of the type of cattle, use as an approximate working guide the slaughter weight/slaughter age graphs in Figure 1.1 to set performance targets. It is much easier

---

* In the EU scheme carcases are classified on a scale 1 (leanest), 2, 3, 4L, 4H, 5L, 5H (fattest). Most buyers prefer fat classes 3 and 4L which correspond to a back fat thickness of about 3 mm and 4 - 5 mm, respectively.

to work with age as a reference now that all cattle in the UK have official passports stating date of birth.

Really the lines on the graph should be broad bands to indicate variation among individual cattle and to recognise that cattle remain in a particular carcase fat class for longer than is commonly supposed. So there is a range of slaughter weights within a stated fat class.

On average, rate of fattening increases as daily gain increases, is faster for early maturing breed types and heifers fatten more rapidly than steers that in turn fatten more rapidly than bulls. So early maturing Hereford cross heifers may traverse a fat class in 40 days at their highest practicable daily gain, whereas Limousin cross bulls fed an all-concentrate diet may take 70 days or even longer at their peak daily gain.

Missing values in Figure 1.1 indicate that a breed type or sex category is not suitable for slaughter at that age because slaughter weight falls outside the range acceptable to buyers.

Breed codes for Figure 1.1: AA x - Aberdeen-Angus cross, BB - Belgian Blue cross, H - Hereford cross, H-F - purebred Holstein-Friesian, LM - Limousin cross, SM - Simmental cross.

The recognised beef systems contributing to Figure 1.1 include:

- Cereal bull beef from dairy-bred calves on an all-concentrate diet with slaughter at around 12 months of age.
- Suckler bull beef on an all-concentrate diet from weaning with slaughter at 12 to 14 months of age.
- Maize silage bull beef with slaughter at about 14 months of age.
- Grass silage bull (or heifer) beef with slaughter at about 14 months of age.
- Eighteen-month beef from autumn-born dairy-bred steers and heifers with slaughter at 16 to 18 months of age.

Figure 1.1 Slaughter age/slaughter weight relationships.

Bulls

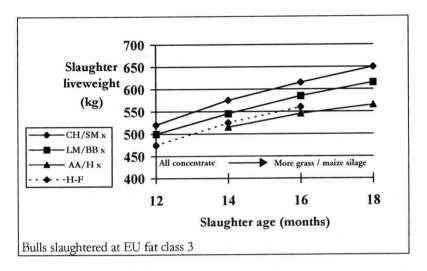

Bulls slaughtered at EU fat class 3

Steers

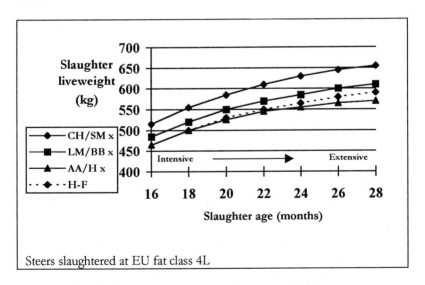

Steers slaughtered at EU fat class 4L

Fig 1.1 (continued)

Heifers

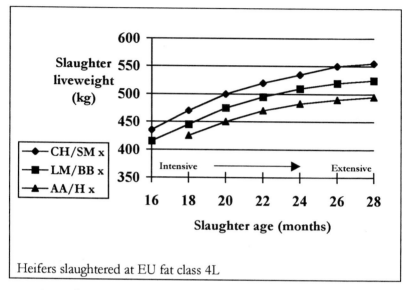

Heifers slaughtered at EU fat class 4L

- Grass finishing of dairy and suckler-bred steers and heifers at 18 to 30 months of age.
- Winter finishing of dairy and suckler-bred steers and heifers at 16 to 30 months of age.

Differences in rate of fattening, not only have important effects on average slaughter weights at preferred fat classes 3 and 4L, but also influence the choice of beef system. The rapid, lean gains of crossbred bulls by continental sires are exploited to greatest advantage in intensive beef systems such as cereal beef. By contrast, early maturing British cross steers and heifers perform well at the lower rates of lifetime daily gain in extensive beef systems with a high reliance on forage. Chapter 7 shows how this knowledge can be exploited in rationing tactics.

The individual variation among cattle within a pen for daily gain and rate of fattening is also extremely important. It is important to weigh and handle cattle to assess individual progress so that cattle are marketed at the right time.

Individual variation also affects marketing tactics. As cattle approach market condition, daily gain declines and feed conversion efficiency worsens. The rate of deterioration is slower than is commonly supposed. Nevertheless, the slow growers that have the poorest feed efficiency soon start losing money. The temptation is to retain them in the vain hope that they will improve. In practice it pays to slaughter slow growers as soon as they enter the buyer's weight range and acceptable fat class to limit the loss. By contrast, fast growers, which are usually slaughtered at the start of the marketing period, continue to grow profitably and should be retained to the top of the buyer's weight range provided they do not become overfat.

The net effect of employing such a policy is usually to shorten the marketing and feeding periods, with the dual advantages of a saving in feed but an increase in average carcase weight.

GROWING CATTLE

For growing cattle due to be grazed in the following summer, winter gains need to be set at a moderate level to exploit compensatory growth - the ability of cattle to gain especially rapidly on high quality grazing after a period of winter feed restriction. This allows important savings in winter feed costs without reducing overall performance.

Guidelines on winter gains are shown in Table 1.1. For summer/autumn-born dairy bred calves the suggested gain of 0.8 kg per day through the rearing winter is a compromise between exploiting compensatory growth whilst ensuring that the calves are sufficiently well grown to make good use of grazed grass when turned out at about six months of age.

For cattle that are to be finished off grass, time comes into the equation as well as compensatory growth. Marketing needs to start in mid grazing season so that cattle numbers are reduced in step with declining grass growth and quality. Also the sale price per kg usually declines from mid-season. So, if cattle are backward at the start of the winter, they should be fed to gain at the upper end of the suggested daily gain range. Similarly, if cattle are being over wintered for sale as stores in the spring, they should be fed at the upper end of the daily gain range, so that they are heavier at sale but still have the lean, hairy appearance that makes a high sale price per kg.

Table 1.1   **Over-wintering gains for cattle to be grazed the following season.**

Category	Daily gain (kg)
Summer/autumn-born dairy-bred calf	
Rearing winter	0.8
Suckled calf/store	
British breed cross	0.4 to 0.6
Continental breed cross	0.6 to 0.8

HEIFER REPLACEMENTS

There are clear-cut advantages in lifetime performance to calving dairy and beef heifers for the first time at around two years of age. However, if autumn-born beef breed x Holstein-Friesian heifers are mated to calve in spring, inevitably they are $2^1/2$ years old at first calving.

Modest daily gains in replacement heifers during rearing are consistent with the highest lifetime performance. Yet in practice heifers often fail to attain even the modest target mating weights, presented in Table 1.2, and are either barren or calve too late in the calving season.

Table 1.2   Target weights for replacement heifers

Breed	Age at calving	Weight at mating	Pre-calving weight
	(years)	(kg)	(kg)
Blue Grey	2	320	485
Continental breed x suckler	2	370	525
Angus/Hereford x			
Holstein-Freisian	2	350	485
	2^1/$_2$	370	520
Continental breed x			
Holstein-Fresian	2	370	525
	2^1/$_2$	390	550

Achieving target weights at low cost depends on good grassland management with the same attention to detail that goes into grass finishing. Heifers, like other cattle, gain best on short leafy swards. Then, as with over wintered stores, rapid compensatory growth during the grazing season can be exploited to save winter feed costs. Winter gains can be set at 0.4 to 0.5 kg per day for Blue Grey and Angus/Hereford x Holstein-Friesian heifers and 0.5 to 0.6 kg for continental breed crosses.

BEEF SUCKLER COWS

The guiding principle in managing beef suckler cows is that body condition is a sensitive barometer of nutritional status. Charting body condition can be used to plan feeding management through the year so that optimum performance is achieved at minimum feed cost.

Body condition is scored on a scale 1 (very thin) to 5 (very fat). The technique is to grip the outer edge of the loin with the thumb curled under the ledge formed by the transverse processes of the spine (Figure 1.2). The ball of the thumb is used to feel the thickness of fat over the bone.

Figure 1.2   The technique of body condition scoring

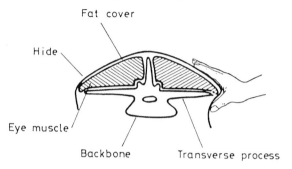

The scores are described as follows, showing the average back fat recorded by Dr Iain Wright of the Macaulay Land Use Research Institute (personal communication):

Score 1 (1 mm back fat)   Bony and emaciated with no detectable fat cover.

Score 2 (4 mm back fat)   The bony ends of the transverse processes can be felt but they are rounded by a thin fat cover.

Score 3 (9 mm back fat)   The transverse processes are covered with fat and can only be felt with firm pressure.

Score 4 (17 mm back fat)   The bone of the transverse processes cannot be felt, even with firm pressure.

Score 5 (28+ mm back fat)   Grossly overfat with substantial puffy fat deposits round the tail head.

In cases of uncertainty, additional guidance can be gained by feeling the fat cover over the ribs with the flat of the hand and pressing with the fingers to feel the amount of fat either side of the tail head. Once the scoring technique has been learned, visual inspection can give a good idea of condition score.

Target condition scores at different stages of the production cycle are shown in Table 1.3. the key score is at mating which should be $2^{1}/2$ for cows mating during the winter and 2 for cows mating on high quality grazing in summer.

Table 1.3   Target condition scores for suckler cows.

Season	Target score	Stage of production	
		Spring calver	Autumn calver
Spring	2	Calving	Pregnant/suckling
Summer	$2^1/_2$	Mating	Weaning
Autumn	3	Weaning	Calving
Winter	$2^1/_2$	Pregnant/dry	Mating

If cows are too thin at mating, conception rates are depressed and re-breeding is delayed with an inevitable and costly increase in the duration of the calving period. Late calvers tend to get later and later until they fail to conceive. And late born calves have below average weaning weights in the autumn with a low valuation. A long calving period also complicates winter feeding management because the cows are at different stages of production and herd rationing underfeeds some, overfeeds others.

Whether cows calve in the spring or autumn, the target condition scores are 3 in the autumn and 2 at turnout in the spring. This allows the loss of up to one unit of condition score over the winter (about 100 kg liveweight) which saves winter feed costs.

The spring calver is in early lactation at turnout but will gain condition to reach a score of at least 2 by mating in mid-summer. At this score the high nutrient intake from grazed grass will be sufficient to allow normal conception. This contrasts with the autumn calver which mates in mid-winter and **must** attain the target score of $2^1/_2$ to conceive readily.

On well managed grazing both spring and autumn calvers should gain condition through the summer and attain a score of 3 by the autumn. If the spring calver exceeds the target score an even greater saving in winter feed may be possible, subject to welfare considerations. However, autumn calvers should be restricted to an autumn score of 3, especially if they are in calf to a heavy continental sire breed, or there is a greater risk of difficult calvings.

If there is a danger of most cows in the herd failing to meet the autumn target, early action must be taken to rectify the situation. Providing plentiful, fresh grazing may suffice. But if necessary, calves should be weaned early to remove the strain of lactation. If only a few cows are at risk, they should be split from the herd and grazed preferentially.

Spring-calving cows can be fed through the winter for a steady loss of condition both before and after calving. For autumn calvers it is wise to allow only a loss of a $1/4$ score by mid winter to support peak lactation whilst ensuring that the target mating score of $2^1/2$ is achieved.

If the target autumn condition score is not met, weight loss through the winter must be restricted. Also, heifers suckling calves on winter rations should not be allowed to lose weight, recognising that they are still developing as well as milking. Even with mature cows it is proposed here that, for the sake of good animal welfare, weight loss should be restricted to not more than 0.5 kg daily.

The milk yield of the cow is determined as much by the appetite of the calf as by yield potential. Milk yield averages around 10 kg per day for the first three months. Thereafter, it declines and on winter rations averages around 7 kg daily during the second half of the winter.

It is important to remember that it is nutritionally more efficient and cost effective to creep feed concentrates directly to the calf than to feed the cow for increased milk production.

Cow weight is determined by breed type and body condition. It is worth weighing at least a sample of cows in the autumn to find out just how much they weigh. Otherwise, the guidelines on autumn weights in Table 1.4 can be used for mature, non-pregnant cows at the target condition score of 3.

Table 1.4   **Autumn weights of commercial adult suckler cows.**

Breed type	Autumn weight at condition score 3 (kg)
Blue Grey	475
Angus cross	575
Hereford cross	575
Belgian Blue cross	625
Limousin cross	600
Simmental cross	600
Charolais cross	675

BREEDING BULLS

During rearing, young breeding bulls need to be offered a high quality diet on which they can exhibit their growth potential, which is a key component of estimated breeding values (EBV) for terminal sire breeds. However, some breeders have pursued high weight gains to extremes without proper regard for the future breeding function of their bulls. The result is that all too many of the bulls offered for sale are too fat for immediate use and there is a risk of structural unsoundness.

The present statistical method of analysing breeding records called Best Linear Unbiased Predictor (BLUP) cleverly separates management influences within and between herds from real genetic effects and produces much more accurate estimates of breeding value than were available formerly. This should remove the temptation to grow bulls faster than is good for them. A genetically superior bull grown at a sensible rate of gain will achieve a higher EBV than a bull of average genetic merit pushed along at peak gain, even though the absolute weight of the superior bull may be lower.

Average daily gain standards to 400 days of age for the principal beef breeds are shown in Table 1.5. Ideally, a ration of a quality able to support these average daily gains would be fed *ad libitum*.

Table 1.5   Daily gain standards for breeding bulls to 400 days of age.

Breed	Daily gain (kg)
Aberdeen-Angus	1.20
Beef Shorthorn	1.20
Belgian Blue	1.30
Blonde d'Aquitane	1.35
Charolais	1.55
Hereford	1.20
Limousin	1.35
Simmental	1.50
South Devon	1.40

A stock bull with a full complement of 30 to 40 cows in a breeding season that may only last around 10 weeks is working hard with several cows in oestrus in a single day. Some loss of body condition is inevitable, especially in young bulls, even if they are hand-fed extra concentrates away from the cows. Therefore, it is important to build up body condition to a score of 3 by the start of the breeding season.

Chapter 2

Feed Intake

The pieces of information needed to devise a ration are the dry matter (DM), metabolisable energy (ME), and crude protein (CP) values of the ration ingredients, ME and protein requirements and feed intake. Feed intake is afforded less attention than it needs and accurate rationing is undermined more by poor predictions of intake than anything else.

AFRC (1993) presented a series of equations for the prediction of dry matter intake (DMI) in lactating, pregnant and growing cattle. Those for rations based on grass silage and fed to growing cattle are not very accurate. Equations for milking cows are not really applicable to suckler cows fed very different rations.

Through the 1990s a good deal of research has been carried out on factors affecting silage intake, notably at the Scottish Agricultural College (SAC) and the Agricultural Research Institute for Northern Ireland (ARINI). Some silage analyses use this research to give an indication of intake potential. At ARINI intakes were recorded in growing cattle for a large variety of grass silage samples collected from commercial farms. The results have been used as a basis for the guidelines on grass silage intakes in Table 2.1 (personal communication from Dr Raymond Steen).

Liveweight, breed type, sex category, body condition, forage quality and the amount of concentrate fed interact to determine feed intake.

GROWING CATTLE

As cattle grow heavier they eat more feed but there is a decline in intake as a percentage of liveweight. On mixed rations of forage and concentrate the traditional guideline is a daily DMI of about 3 per cent of liveweight at 100 kg, falling to 2.5 per cent at

300 kg and 2 per cent or less at 600 kg. It is very unreliable as a predictor of feed intake.

Breed and sex effects on daily gain are related in part to differences in feed intake. Continental breed types may consume 10 per cent more feed than British crosses at the same liveweight and bulls 10 per cent more than steers. The important point is to recognise the differences in daily gain potential of the various categories of cattle. These and differences in feed efficiency are built into the ME allowances in Chapter 3.

Cattle in plain body condition after a period of underfeeding, when fed high quality rations eat more to fuel compensatory growth than those that are already in good condition. The difference in intake can be 10 to 15 per cent, which is sufficiently large to be taken into account when predicting feed intake.

Hay is eaten in larger quantities than grass of the same quality conserved as silage. Intake is directly proportional to ME value, subject to the hay being well made.

Straws with their low ME and CP values are eaten in smaller quantities even than poor quality hay. Upgrading the nutritional value of straw, for example by stack treatment with ammonia, increases intake potential.

The intake of silage is especially difficult to predict since fermentation quality and DM content influence intake, as well as ME and CP values. Key factors are the liveweight of cattle, the depression of forage intake when concentrates are fed (concentrate substitution), the DM of the silage, its digestibility (ME), CP content and its fermentation quality as measured in part by its ammonia nitrogen content (high values poorer).

Wilted silage with a high DM content of 25 per cent is eaten in larger quantities than wetter silage at 20 per cent DM or less. Short chopped silage has a higher intake than long material. This becomes especially important if silage is self-fed from a clamp and for big bales because the sheer compaction of the silage makes feeding hard work for the cattle.

Forage feeds are commonly supplemented with concentrates which partially substitute for forage intake. This is called concentrate substitution. There is an increase in total DMI even though forage intake is depressed. Feeding trials show depressions in grass silage intake of 0.4 to 0.65 kg DM per kg concentrate DM fed. However, it should be noted that at very high levels of concentrate feeding (over 80 per cent of the ration DM) concentrate substitution can approach 1 kg reduction in silage DMI per kg concentrate DM fed.

It is logical to expect that, the poorer the forage quality, the lower will be the rate of concentrate substitution. This proves to be the case and straw DMI may fall only 0.2 kg per kg concentrate DM fed. Indeed straw intake may actually increase when concentrates are fed in response the additional protein supplied.

As the level of concentrates increases towards *ad libitum*, bulk restrictions on intake recede and the limitation is energy intake. So, for example, cereal beef cattle fed an all-concentrate diet may consume less DM than cattle of similar weight fed a mixed forage/concentrate ration. With high quality maize silage it seems that, up to 30 per cent of the ration DM as silage, energy status controls feed intake and daily gain in cattle is similar to those fed all-concentrates. Thereafter, concentrate substitution comes into play.

Guidelines on feed intake for growing cattle in good body condition are presented in Table 2.1. Forage intakes are potential intakes if the forage is fed *ad libitum* as the sole diet and for grass silages are for short-chopped material. Adjustments to the base values are suggested for long unchopped material and for cattle in plain body condition. It is wise to err on the side of caution in predicting forage intake rather than be over-optimistic and find in mid-winter that the cattle have fallen way behind target

Table 2.1. Guidelines on potential forage dry matter (DM) intakes when fed as the sole diet to growing cattle in good body condition. Source: Author estimates

Liveweight (kg)	Grass silage Good[1]	Grass silage Moderate[2]	Big bale silage 10.5 ME	Maize silage 11 ME	Hay 10 ME	Hay 9 ME	Straw 7.5 ME[3]	Straw 6.5 ME[4]	Concentrate 12.5 ME
				Feed intake (kg DM per day)					
200	4.3	3.6	3.9	4.5	4.0	3.6	2.7	2.4	6.0
300	5.8	4.8	5.2	6.1	5.4	4.8	3.6	3.2	8.0
400	7.2	6.0	6.5	7.6	6.7	6.0	4.5	4.0	9.0
500	8.5	7.1	7.7	9.0	7.9	7.1	5.3	4.8	9.5
600	9.7	8.1	8.8	10.3	9.1	8.1	6.1	5.5	10.0
				Reduction of forage DM intake per concentrate DM fed[5]					
	0.5	0.4	0.5	0.6	0.5	0.4	0.3	0.2	-

1 Good quality silage 25% DM, 11 MJ ME per kg DM with good fermentation characteristics
2 Moderate quality silage with 20% DM, 10 MJ ME per kg DM with poor fermentation characteristics
3 Ammonia treated barley straw
4 Untreated barley straw
5 Reduction in forage DM per kg concentrates fed at medium levels of concentrate feeding. At very high levels of concentrate feeding substitution can approach 1kg reduction in silage DMI per concentrate DM fed.

Adjustments

Unchopped grass silage **deduct** 5 per cent
Grass silage less than 20 per cent DM **deduct** 5 per cent
Cattle in plain body condition **add** 10 per cent

performance because they were unable to consume the ration.

In the rationing procedure described in Chapter 6, the predicted intakes of forage are used as a measure of forage potential in rations that maximise forage content. The ME that would be provided by forage alone is compared with the ME requirement and any shortfall is calculated. The shortfall is then divided by the net value of 1 kg concentrates taking account of substitution, ie ME of 1 kg concentrate DM minus the ME value of reduced forage DM intake. This figure indicates the level of concentrate DM required to support the target gain. Then the actual daily allowance of silage can be calculated from the total ME requirement minus that provided by concentrates.

Forage intakes are low in reared dairy-bred calves at around three months of age when they weigh only around 100 kg. Forage can be introduced into the diet gradually from eight to ten weeks of age but the daily concentrate allowance should be at least $2^1/_2$ kg.

SUCKLER COWS

The principles outlined for growing cattle apply similarly to the prediction of feed intake in suckler cows. However, the lactating cow with her urge to milk, even at the cost of losing weight and body condition, has a greater feed intake capacity than dry cows that are little more than idling nutritionally. There is no reason to suppose that dry cows consume much more feed than growing cattle of similar weight, though they may handle low quality forages better.

Guidelines on feed intakes of suckler cows are shown in Table 2.2, together with concentrate substitution rates. Intakes for lactating cows are set 15 per cent above dry cows. Even at peak lactation on high quality forages, appetite limits rarely come into play when formulating rations. On the contrary, it is more often the case that the ration needs to be 'diluted' with straw fed to appetite to avoid cows over-eating more expensive ration ingredients.

Table 2.2. Guidelines on potential forage dry matter (DM) intakes when fed alone to suckler cows. Source: Author estimates

Liveweight (kg)	Grass silage Good[1]	Grass silage Moderate[2]	Big bale silage 10.5 ME	Hay 10 ME	Hay 9 ME	Straw 7.5 ME[3]	Straw 6.5 ME[4]
				Feed intake (kg DM per day)			
Lactating cows							
500	10.3	8.5	9.1	9.7	8.6	6.7	6.1
600	11.9	9.8	10.5	11.2	9.9	7.7	7.0
700	13.3	11.0	11.7	12.5	11.1	8.6	7.8
Dry cows							
500	9.0	7.4	7.9	8.5	7.5	5.8	5.3
600	10.3	8.5	9.1	9.7	8.6	6.7	6.1
700	11.6	9.5	10.2	10.9	9.7	7.5	6.8
			Reduction of forage DM intake per concentrate DM fed[5]				
	0.5	0.4	0.4	0.5	0.4	0.3	0.2

1 Good quality silage 25% DM, 11 MJ ME per kg DM with good fermentation characteristics
2 Moderate quality silage with 20% DM, 10 MJ ME per kg DM with poor fermentation characteristics
3 Ammonia treated barley straw
4 Untreated barley straw
5 Reduction in forage DM per kg concentrates fed at medium levels of concentrate feeding. At very high levels of concentrate feeding substitution can approach 1kg reduction in silage DMI per concentrate DM fed.

Adjustments

Unchopped grass silage **deduct** 5 per cent
Grass silage less than 20 per cent DM **deduct** 5 per cent

Chapter 3
ME Requirements

METABOLISABLE ENERGY

Metabolisable energy (ME) is the unit of energy used in the UK ruminant rationing system and many others world wide. It is the energy circulating in the bloodstream after losses from the gross energy of the feed from indigestible faecal energy, methane gas from the rumen and energy in the urine (Figure 3.1). Work done on the ME preparing it for use in body functions results in a further loss of energy called the heat increment. The net energy that is left is available for body maintenance, physical activity, growth and milk production.

Using ME as the basis of a rationing system for cattle depends on evaluating the ME contents of feed ingredients, expressed as megajoules (MJ) of ME per kg DM, and establishing daily ME requirements to support target levels of performance.

Figure 3.1 Partition of feed energy.

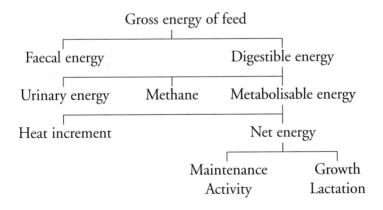

FEED TABLES

Table 3.1 shows typical ME contents of selected feeds expressed as MJ per kg DM; this is also referred to as the M/D of the feed. The table also shows percentage values for the digestibility ('D') of organic matter in feeds.

If only the 'D' value of a feed is known an approximate measure of ME can be calculated by multiplying 'D' (%) by 0.15, e.g.:

$$70 \text{ 'D' } \times 0.15 = 10.5 \text{ MJ ME per kg DM.}$$

Table 3.1 also presents other measures of nutritional quality. Dry matter and crude protein (CP) are presented as decimal values of 1 rather than the more conventional per cent or g per kg DM[1]. This is done because it is how the values are used when calculating a ration.

For major dietary ingredients such as silage or hay that vary widely in quality, a feed analysis is good value for money. Cereals also vary in quality sufficiently to justify analysis. Consultants and feed companies are usually able to provide feed analyses from approved laboratories.

---

[1] Percentages of 10 and over or 100g per kg or more can be converted to decimal proportions simply by placing a decimal point in front of the digits, eg 15 per cent = 0.15 and 125g per kg = 0.125. Values less than 10 per cent or 100g per kg need a zero before the decimal point, eg 7 per cent - 0.07 and 75g per kg = 0.075

Table 3.1   Nutritive value of selected feeds

Feed	DM (Decimal)	"D" (%)	ME (MJ/kg DM)	CP (Decimal)	Degradability[1] (Low-med-high)	Ca	P (g per kg DM)	Mg
Silage								
High quality	0.22	70	11.0	0.165	H	6.4	3.0	1.5
Good quality	0.22	67	10.5	0.160	H	6.8	3.1	1.6
Medium quality	0.22	63	10.0	0.155	H	6.5	3.0	1.4
Low quality	0.22	57	9.0	0.150	H	6.2	2.9	1.3
Maize silage	0.30	75	11.3	0.088	H	3.8	2.7	1.3
Whole crop wheat + urea	0.55	65	10.5	0.240	H	2.4	1.8	0.9
Cereals								
Barley	0.86	86	13.3	0.129	H	1.1	3.9	1.2
Maize	0.86	87	13.8	0.102	M	0.1	3.0	1.3
Oats	0.86	68	12.0	0.108	M	0.8	3.4	1.0
Wheat	0.86	87	13.6	0.130	H	0.4	3.4	1.1
Hay								
High quality	0.85	67	10.5	0.130	H	6.0	2.9	1.6
Medium quality	0.85	63	9.5	0.118	H	5.3	2.6	1.4
Low quality	0.85	57	8.0	0.085	H	5.0	2.0	1.5

Feed	DM (Decimal)	"D" (%)	ME (MJ/kg DM)	CP (Decimal)	Degradability[1] (Low-med-high)	Ca	P (g per kg DM)	Mg
Straw								
Barley untreated	0.86	44	6.5	0.040	-	4.5	0.9	0.8
-NH₃[2]	0.86	52	7.5	0.090	-	4.6	0.8	0.6
Oat	0.86	46	7.2	0.034	-	2.4	0.9	0.8
Wheat	0.86	40	6.1	0.039	-	3.8	0.7	0.8
Protein sources								
Fishmeal (white)[2]	0.90	80	14.2	0.701	L	79.3	43.7	2.2
Protein conc[3]	0.87	75	12.0	0.330	M	18.0	8.0	4.0
Rapeseed meal[4]	0.90	75	13.3	0.400	H	7.8	12.0	4.5
Soyabean meal	0.90	80	13.4	0.505	M	4.5	7.6	2.9
Miscellaneous								
Distillers dark grains	0.90	80	13.0	0.290	M	0.5	3.0	1.5
Draff	0.23	58	10.5	0.200	M	1.0	3.5	1.0
Flaked maize	0.90	90	14.5	0.110	M	0.1	3.0	1.3
Fresh brewer's grains	0.28	60	11.5	0.220	M	3.3	4.1	1.5
Maize gluten	0.90	80	12.2	0.230	M	3.0	9.0	4.3
Molassed sugar beet pulp	0.87	82	12.5	0.111	M	5.9	0.7	1.0

Feed	DM (Decimal)	"D" (%)	ME (MJ/kg DM)	CP (Decimal)	Degradability[1] (Low-med-high)	Ca	P (g per kg DM)	Mg
Roots								
Carrots	0.11	81	12.8	0.085	H	5.9	3.4	1.8
Fodder beet	0.18	70	11.9	0.070	H	3.9	1.8	1.4
Mangels	0.11	79	12.4	0.085	H	3.9	1.8	1.4
Potatoes	0.21	79	13.3	0.115	H	0.4	2.0	1.0
Swedes	0.11	82	14.0	0.095	H	3.5	2.6	1.1
Turnips	0.10	72	12.7	0.100	H	4.8	3.4	1.1

1   L = low; M = medium; H = highly degradable
2   The EU has imposed a temporaray ban on fishmeal as a BSE control measure from 1 January 2001.
3   NH_3 = straw treated with ammonia
4   Typical proprietary protein concentrate
5   Extracted

THE ME SYSTEM

A problem in using ME as the basis of an energy rationing system is that the efficiency of utilisation of ME varies according to the ME concentration (M/D) of the ration. This creates a "chicken and egg" situation in which, to estimate daily gain in growing cattle, it is necessary to know the M/D of the ration but, to calculate the M/D required, it is necessary to know the daily gain. The same situation applies to suckler cows.

The conventional solution to this apparent impasse is to calculate energy requirements as net energy values and then work back to a convenient set of ME allowances for cattle at different levels of performance fed rations of varying M/D concentrations. A computer handles the calculations with ease but for hand calculation the original ME system for growing cattle used a set of tabulated values that made ration formulation very tedious.

The whole process can be simplified by starting the calculation with a prediction of what the final M/D of the ration is expected to be. That more or less disposes of the interaction between ration quality and performance. If the actual M/D of the ration differs by more than, say, 0.5 from the predicted value, it is simple enough to make a new prediction of M/D and adjust the ration. This novel approach is sufficiently accurate for practical ration formulation and is used here.

Sometimes the actual ration M/D is known, as with an all-concentrate diet or a total mixed ration (TMR). If not, for mixed diets that maximise forage intake, predicted overall ration M/D values for suckler cows are low in relation to the feeds commonly used on farms. Often the forage has a higher M/D than the Table 3.2 value and cows are rationed below appetite or fed limited silage with straw to appetite. In this case use the table value.

Table 3.2   Prediction of final rotion M/D at various levels of performance.

(a) Growing/finishing cattle

Cattle type	A	B	C	D	E
Daily gain (kg)	**Predicted ration M/D** (MJ ME/kg DM)				
0.6	na	10.6	10.7	10.8	10.9
0.8	na	10.8	10.9	11.0	11.1
1.0	10.8	11.0	11.2	11.3	11.4
1.2	11.1	11.3	11.5	11.6	na
1.4	11.5	11.7	11.9	na	na
1.6	12.0	na	na	na	na

na - gain not applicable to this type of cattle

Cattle type A: Continental breed or crossbred bulls
             B: Continental breed or crossbred bulls
                 British breed or crossbred bulls
             C: Continental breed or crossbred heifers
                 British breed or crossbred steers
                 Holstein-Friesian bulls
             D: British breed or crossbreed heifers
                 Holstein-Friesian steers
             E: Holstein-Friesian heifers

(b) Suckler cows

	Predicted M/D (MJ ME/kg DM)
Early lactation	10.5
Late lactation	9.5
Dry, pregnant	8.5

GROWING CATTLE

The allowances proposed in this manual are based on the AFRC (1990, 1993) proposals but with some important modifications outlined below and detailed in Appendix 1:

- The AFRC classification of breeds into early, medium and late maturing types has not been accepted. The preferred classification is:

  Continental breeds and crosses (most efficient)

  British breeds and crosses

  Holstein-Friesian (least efficient)

- AFRC assumed that more active bulls have a maintenance energy requirement 15 per cent higher than steers or heifers. For daily gain bulls are assumed to have an energy value of gain 15 per cent lower than steers which, in turn, have a value 15 per cent lower than fatter heifers.

  In this manual, housed bulls, which effectively covers all bull beef production in the UK, are assumed to have the same maintenance requirement as steers and heifers. Sex effects on the energy value of gains have been reduced to a 10 per cent lower value for bulls than steers and 10 per cent lower for steers than heifers.

Taken together, the breed and sex effects on energy requirements adopted here narrow the range of ME allowances for a stated daily gain compared to AFRC (1990, 1993). The ME allowances presented in Table 3.3 are divided into five categories:

- Continental breed and crossbred bulls.
- Continental breed and crossbred steers. British breed and crossbred bulls.
- Continental breed and crossbred heifers. British breed and crossbred steers.

- Holstein-Friesian bulls.
- British breed and crossbred heifers. Holstein-Friesian steers.
- Holstein-Friesian heifers.

The reason why five sections are necessary is simply that the variation in ME requirements for the different categories of cattle demand them. For example, in 400 kg cattle fed an 11 M/D ration to gain 0.9 kg per day, ME requirements vary from 78 MJ ME for continental breed bulls to 99 MJ ME for Holstein-Friesian heifers. The consolation is that it is only necessary to consult one of the sections for a particular category of cattle.

ME allowances are presented for the practical range of M/D values, cattle weights and daily gains. Missing values at higher rates of daily gain denote that they are probably beyond genetic potential. Missing values at lower rates of gain indicate that the DM intake would be so far below appetite that cattle welfare would be compromised.

The ideal ration is one that provides the required ME intake whilst satisfying appetite. If the ration M/D provides too much ME, though satisfying appetite, it is simple enough to reduce M/D, for example by feeding straw, so that ME intake and appetite are in balance.

As daily gains increase at the upper end of the range, note that there is an increase in the ME needed to produce an extra 0.1 kg daily gain. This is most pronounced for early maturing cattle types such as British breed cross heifers and reflects the higher energy content of gains in fatter cattle. It does not contravene the general principle that the overall efficiency of utilisation of ME per kg gain improves as daily gain increases because maintenance is a lower proportion of the total ME requirement.

There are situations where the ME requirement may be lower than the tabulated value. The classic case is where cattle fed a moderate quality store ration are then transferred to a high quality feed (usually spring grazing). The lean compensatory growth utilises ME more efficiently than the same daily gain in cattle fed continuously on a high quality ration.

Compensatory growth can play a part in winter finishing. Grazed calves and stores that have achieved only moderate gains during grazing may have the potential for compensatory growth in winter, whereas fast-grown suckled calves do not. This is not a reason to restrict grazing gain, maximising which is essential for profitable production in beef systems employing grazing.

Another situation in which efficiency of utilisation of ME is improved is where growth promoters are used. The effect is greatest for hormone implants because, not only is daily gain increased, but also the gains are leaner. All hormone implants are banned in the EU but some products are still licensed in some countries, e.g. the USA.

In-feed growth promoters, or digestive enhancers as they are sometimes called, improve feed conversion efficiency by 10 per cent and this may or may not be associated with increased daily gain. Compounds such as monensin sodium act selectively on the rumen microorganisms to reduce waste methane production and to shift fermentation favourably towards a higher proportion of propionic acid in the rumen mix of volatile fatty acids that ruminants use, in part, as a source of energy. The future of these products is uncertain in the EU and some company buying specifications already exclude them.

At the other end of the scale, gains and feed conversion efficiency may be sub-normal if cattle are clinically or sub clinically unhealthy. For example, at the end of the grazing season there is often a burden of inhibited parasitic worm larvae in the gut lining waiting to develop and cause scouring and reduced daily gain later in the winter. In this event it is cost effective to treat cattle at yarding with an effective anthelmintic. With some products there is the added advantage that external parasites are also controlled.

Table 3.3 ME requirements of growing cattle.
(Adapted from AFRC, 1990)

(a)   Continental breed/cross bulls

Live-weight (kg)	Ration M/D	MJ ME per day at a daily gain (kg) of:								
		0.8	0.9	1.0	1.1	1.2	1.3	1.4	1.5	1.6
100	11	31	33	36	-	-	-	-	-	-
	12	29	31	33	36	38	41	-	-	-
	13	28	30	32	34	36	38	41	44	-
200	10	49	53	-	-	-	-	-	-	-
	11	46	50	53	57	61	65	-	-	-
	12	44	47	50	53	57	61	65	69	74
	13	-	-	-	-	54	57	60	64	69
300	10	64	68	73	-	-	-	-	-	-
	11	61	64	69	73	78	84	90	-	-
	12	-	61	65	69	74	78	83	89	95
	13	-	-	-	-	69	74	78	83	88
400	10	77	83	89	-	-	-	-	-	-
	11	74	78	83	89	95	102	109	-	-
	12	-	73	78	84	95	89	101	108	115
	13	-	-	-	-	-	89	95	100	107
500	10	90	96	103	-	-	-	-	-	-
	11	86	91	97	103	110	118	126	-	-
	12	-	87	92	97	103	110	117	125	134
	13	-	-	-	-	-	104	110	116	124

Note: Missing values at higher gains above potential and at lower gains below acceptable DM allowance.

(b) Continental breed/cross steers and British breed/cross bulls

Live-weight (kg)	Ration M/D	0.6	0.7	0.8	0.9	1.0	1.1	1.2	1.3	1.4	1.5
		MJ ME per day at a daily gain (kg) of:									
100	10	29	32	-	-	-	-	-	-	-	-
	11	28	30	33	35	-	-	-	-	-	-
	12	27	29	31	33	36	39	42	-	-	-
	13	-	27	29	31	34	36	39	42	-	-
200	10	45	48	52	56	-	-	-	-	-	-
	11	43	46	49	53	57	61	66	71	-	-
	12	-	44	47	50	53	57	61	66	71	-
	13	-	-	-	47	51	54	58	62	66	70
300	9	62	-	-	-	-	-	-	-	-	-
	10	58	63	68	73	-	-	-	-	-	-
	11	56	60	64	69	74	79	85	92	-	-
	12	-	-	61	65	69	74	79	85	91	98
	13	-	-	-	-	-	70	75	79	85	91
400	9	75	-	-	-	-	-	-	-	-	-
	10	71	76	82	88	-	-	-	-	-	-
	11	68	73	78	83	89	96	103	111	-	-
	12	-	-	74	79	84	90	96	103	110	119
	13	-	-	-	-	-	-	90	96	102	109
500	9	87	-	-	-	-	-	-	-	-	-
	10	83	89	95	102	-	-	-	-	-	-
	11	79	84	90	97	104	111	119	129	-	-
	12	-	-	86	92	98	104	111	119	128	137
	13	-	-	-	-	-	-	105	112	119	127

Note: Missing values at higher gains above potential and at lower gains below acceptable DM allowance.

(c) Continental breed/cross heifers, British breed/cross steers and Holstein/Friesian bulls

Live-weight (kg)	Ration M/D	MJ ME per day at a daily gain (kg) of:										
		0.5	0.6	0.7	0.8	0.9	1.0	1.1	1.2	1.3	1.4	1.5
100	10	28	31	-	-	-	-	-	-	-	-	-
	11	27	29	32	35	38	-	-	-	-	-	-
	12	26	28	30	33	35	39	42	-	-	-	-
	13	-	27	29	31	33	36	39	42	-	-	-
200	10	43	47	-	-	-	-	-	-	-	-	-
	11	41	45	48	52	56	61	66	-	-	-	-
	12	-	-	-	49	53	57	61	66	72	78	-
	13	-	-	-	-	50	54	58	62	66	71	77
300	9	59	65	-	-	-	-	-	-	-	-	-
	10	56	61	66	72	-	-	-	-	-	-	-
	11	54	58	63	67	73	79	85	93	-	-	-
	12	-	-	60	64	69	74	79	86	92	100	-
	13	-	-	-	-	-	70	75	80	86	92	99
400	9	72	78	-	-	-	-	-	-	-	-	-
	10	69	74	80	87	-	-	-	-	-	-	-
	11	65	71	76	82	88	95	103	112	-	-	-
	12	-	-	72	78	83	89	96	103	111	120	130
	13	-	-	-	-	-	-	90	97	103	111	119
500	9	84	91	-	-	-	-	-	-	-	-	-
	10	80	86	93	102	-	-	-	-	-	-	-
	11	76	82	88	95	103	111	120	130	-	-	-
	12	-	-	84	90	97	104	111	120	129	140	151
	13	-	-	-	-	-	-	105	112	120	129	138

Note: Missing values at higher gains above potential and at lower gains below acceptable DM allowance.

(d) British breed/cross heifers and Holstein/Friesian steers

Live weight (kg)	Ration M/D	MJ ME per day at a daily gain (kg) of:									
		0.5	0.6	0.7	0.8	0.9	1.0	1.1	1.2	1.3	1.4
100	10	29	32	-	-	-	-	-	-	-	-
	11	28	31	34	37	-	-	-	-	-	-
	12	27	29	32	35	38	41	-	-	-	-
	13	-	28	30	33	35	39	42	46	-	-
200	10	45	49	-	-	-	-	-	-	-	-
	11	43	46	50	55	60	65	-	-	-	-
	12	-	44	48	52	56	61	66	72	78	-
	13	-	-	-	-	53	57	61	66	72	78
300	10	59	64	70	76	-	-	-	-	-	-
	11	56	61	66	71	77	84	92	-	-	-
	12	-	-	62	67	73	78	85	92	100	-
	13	-	-	-	-	69	74	79	85	92	99
400	10	71	77	84	92	-	-	-	-	-	-
	11	68	73	80	86	94	102	111	-	-	-
	12	-	70	76	81	88	95	103	111	121	-
	13	-	-	-	-	-	89	96	103	111	120
500	10	83	90	98	107	-	-	-	-	-	-
	11	79	86	93	100	109	118	129	-	-	-
	12	-	82	88	95	102	110	119	129	140	-
	13	-	-	-	-	-	104	111	120	129	139

Note: Missing values at higher gains above potential and at lower gains below acceptable DM allowance.

(e)   Holstein/Friesian heifers

Live-weight (kg)	Ration M/D	MJ ME per day at a daily gain (kg) of:							
		0.5	0.6	0.7	0.8	0.9	1.0	1.1	1.2
100	11	29	32	35	-	-	-	-	-
	12	28	30	33	37	40	44	-	-
	13	26	29	32	34	38	41	45	49
200	10	46	51	56	-	-	-	-	-
	11	44	48	53	58	64	70	-	-
	12	-	46	50	54	59	65	71	77
	13	-	-	-	51	56	60	65	71
300	10	61	67	73	-	-	-	-	-
	11	58	63	69	75	82	90	-	-
	12	-	60	65	71	77	83	91	99
	13	-	-	-	-	72	78	84	91
400	10	74	81	89	-	-	-	-	-
	11	70	76	83	91	99	109	-	-
	12	-	73	79	85	93	101	110	120
	13	-	-	-	-	87	94	102	110
500	10	86	94	103	-	-	-	-	-
	11	82	89	97	106	115	126	-	-
	12	-	85	92	99	108	117	127	139
	13	-	-	-	-	102	110	118	128

Note: Missing values at higher gains above potential and at lower gains below acceptable DM allowance.

SUCKLER COWS

ME allowances for housed suckler cows are based on AFRC (1990, 1993) proposals for dairy cows with only minor modifications (see Appendix 2). For autumn calvers, mating is assumed to be in the middle of a six-month winter. Spring calvers are assumed to have a two-month suckling period on winter rations before they are turned out to graze.

The ME requirements in Table 3.4 show several missing values at a low M/D value of 8 to indicate probable appetite limits and missing values at a high M/D of 11 to indicate that ration quality would be too high for the performance level. Nevertheless, high ME silage can be fed provided it is diluted, for example with straw, to allow cattle to eat their fill at the required ME intake.

The proposed weight losses in Table 3.4 are for cows attaining the target condition score of 3 in the autumn. Feed costs are saved by allowing the cows to lose weight and condition through the winter ("feed off their backs"), so that they are turned out to graze at a condition score of 2. The loss of a unit of condition score represents a loss of 90 to 100 kg liveweight.

All too often, cows fail to achieve the target autumn condition score of 3 in the autumn. If, for example, the autumn condition score is only $2^{1}/2$ the permissible weight loss over the winter is reduced to 45 to 50 kg. Table 3.4 shows that an extra 20 MJ ME per day must be fed if no weight loss can be allowed. In any event, care must be taken to ensure that autumn calvers do not fall below the critical condition score at mating in mid-winter of $2^{1}/2$. So a modest weight loss of 0.25 kg per day is proposed between calving and mating. If there is a time to feed suckler cows generously, this is it.

In theory even greater weight losses are possible over winter if cows exceed condition score 3 in the autumn after a good grazing season. In practice there are limits to weight loss for two reasons. Firstly, whilst cows can "milk off their backs" there is a danger of reduce milk yield if weight loss is too rapid. And secondly, the weight loss should not contravene cattle welfare standards. On both counts the conclusion is that weight loss should not exceed 0.5 kg per day.

The first-calved heifer is under pressure to produce milk whilst still developing towards maturity, so she should not be allowed to lose weight through the winter.

Table 3.4   **ME allowances for housed suckler cows.**
Adapted from AFRC (1990, 1993)

Liveweight (kg)	Weight loss (kg/day)	Milk yield(kg)	M/D (MJ ME/kg DM) 8	9	10	11
SPRING CALVER						
Pre-calving [1]			ME requirement (MJ/day)			
500	-0.50	0	58	57	55	-
600			66	64	62	-
700			73	71	70	-
Post-calving [1]						
500	-0.50	10	-	92	89	87
600			-	100	97	94
700			-	108	105	102
AUTUMN CALVER						
Pre-mating [2]						
500	-0.25	10	-	102	99	96
600			-	110	107	104
700			-	118	114	111
Post-mating [1]						
500	-0.50	7	78	76	74	-
600			86	84	81	-
700			94	92	89	-

[1] If no weight loss permissible **ADD** 20 MJ ME/day
[2] If no weight loss permissible **ADD** 10 MJ ME/day

Chapter 4

Protein Requirements

Energy rightly holds centre stage in rationing because it is the driving force for growth. Yet it is dietary protein that provides the building blocks for lean meat production. Moreover, unless the energy in the diet is correctly balanced with protein, minerals and vitamins, utilisation of the whole ration is impaired.

PROTEIN DEGRADABILITY

Protein nutrition is more complicated in beef cattle than in simple-stomached animals such as the pig because of the part played by the rumen in protein utilisation. The microorganisms in the rumen have a protein requirement of their own which must be met by effective rumen degradable protein (ERDP) which replaces the former term rumen degradable protein (RDP). The dietary protein is first degraded to ammonia and then built into microbial protein.

It follows that the microorganisms also have the important capacity to turn non-protein nitrogen (NPN) sources such as urea into microbial protein. To do this they need to be provided with an energy source to fuel their metabolism and a small quantity of sulphur which is incorporated into some of the amino acids of microbial protein.

Not all feed components are fermented in the rumen and some of those that are do not supply energy to the microorganisms, though they provide nutrients to the host animal when digested in the lower gut. In particular, fats and protein itself yield negligible energy to the microorganisms and the organic acids in silage and ensiled brewers grains yield none at all.

The most important sources of fermentable energy for the rumen microorganisms are starches, sugars, celluloses and hemicelluloses.

Not only are they important sources of energy for the host animal, but they also have an important influence on protein utilisation. The yield of fermentable energy from a feed is known as its fermentable metabolisable energy (FME). The proportion of FME in the ME varies from 0.70 in poor butyric silage to 0.95 in a cereal concentrate.

Microbial protein eventually passes out of the rumen and is digested in the abomasum (stomach) and small intestine by protein enzymes.

Situations can arise when there is insufficient microbial protein to supply the host animal's total requirement. This is particularly so for young cattle growing rapidly which have a high proportion of lean in their gain. The additional requirement must be supplied as less soluble dietary protein which escapes rumen breakdown and is digested in the lower gut. This fraction is called digestible undegraded protein (DUP) instead of the former term undegraded dietary protein (UDP).

So the way in which protein is digested can be described by its degradability. Table 3.1 (Chapter 3) shows the overall CP content of feeds together with an approximate rating for degradability, which is classified as high (70+ per cent degradable), medium, or low (under 50 per cent degradable).

Fishmeal is the classic example of a protein source which combines low degradability with high digestibility in the lower gut. At the other end of the scale, silage with its high content of NPN is highly degradable and, for young cattle, must be supplemented with DUP.

It would be wrong to assume that DUP matters more than ERDP. This is not so. Beyond six months of age DUP requirements are relatively low and can easily be met by mixed diets. However, if ERDP is deficient in the ration, the growth of microorganisms is restricted, energy digestion in the rumen is impaired and forage intake is reduced. Beyond calfhood ERDP is most likely to be the limiting protein factor in beef cattle rations.

Equally, with the highly degradable protein of silage, energy may limit utilisation of the NPN fraction. This is often a bigger problem than might be supposed because about 10 per cent of the ME of unwilted silage is as preformed fatty acids and is not available as FME to the microorganisms. The combination of highly degradable protein and low FME means that CP calculated by the convention of N content x 6.25 grossly overvalues some silages as protein sources.

For the same reason it pays to feed diets containing NPN materials, such as urea, little and often and to include a readily fermented source of energy such as cereals, to assist the utilisation of N.

At the other extreme high DUP feeds such as fishmeal, which combine low degradability with a low proportion of FME, can leave the rumen microorganisms short of ERDP.

Fishmeal, fed at 0.25kg per day, has the particular capability to bring about an increase in silage consumption by cattle beyond the age DUP is a limiting factor. The response is associated with improved fibre digestion. It is not clear whether protein nutrition is in some way involved in the response.

PROTEIN RATIONING SYSTEMS

The principles of protein utilisation have been incorporated into working party proposals for a metabolisable protein (MP) system to be a companion to the ME system (AFRC, 1992. 1993).

The MP system for beef cattle is still being evaluated and has not been adopted here. The general conclusion can be drawn from the working party report that, in practical beef cattle rations, the main protein requirement is for ERDP. Provided care is taken in selecting and combining ration ingredients, CP is acceptable as a basis for rationing.

As seen earlier, over the range of feed ME concentrations from 9 MJ/kg DM (poor silage) to 13 (high quality concentrates), the proportion of the ME present as FME rises from 0.70 to 0.90. The requirement for ERDP has been set by AFRC (1992) at 10g per MJ FME and therefore rises as ration M/D increases. Beyond calfhood the main need is for highly degradable protein sources to supply enough ERDP to match the FME, and so it follows that for beef cattle there is a strong correlation between the required CP proportion in the diet and its M/D.

Whilst CP is used for protein rationing in this manual, the recommended protein contents of rations have been deduced from calculations using the new MP system (Alderman, personal communication).

As a general rule, when selecting ingredients the aim should be a ration which avoids extremes of degradability. Where ingredients with highly degradable protein form the basis of the ration, a source of fermentable energy should be included to aid utilisation. Examples of feed ingredients which need careful balancing:

Grass silage is often overvalued as a source of CP. Much of the protein is highly degradable and this is confounded by the low content of FME.

Maize silage has a high FME but low CP content. It requires supplementary ERDP, with urea as a possible source for cattle over six months old.

Urea in totally degraded in the rumen and provides no energy. It is poorly utilised by cattle under six months, is toxic in excess and is usually limited to not more than 1 per cent of the total diet.

Whole-crop, cereal + urea is high in degradable protein and high in FME.

Roots and cereals contain highly degradable protein and are high in FME.

Fishmeal has a very high protein content of low degradability. It is subject to a temporary ban in the EU from January 2001.

Rapeseed, soya bean meal and field beans are high in protein which is highly degradable. They are also high in FME.

By-products need particular care in checking protein degradability and FME content.

PROTEIN REQUIREMENTS

Protein requirements are expressed in this manual as the CP proportion needed in the DM to support the target performance. Values are expressed as decimal proportions[1] rather than percent or g per kg DM because this is the figure used to calculate the ration.

[1] 15% = 0.15 and 125g/kg = 0.125; 7% = 0.07 and 75g/kg = 0.075.

GROWING CATTLE

Table 4.1 shows the CP proportions in the ration DM required by growing cattle. It may come as a surprise that there is no mention of daily gain. CP content is related only to ration M/D and liveweight. Liveweights above 200kg are not differentiated, nor is cattle type.

Table 4.1 Recommended crude protein contents in diets for growing cattle (Alderman, personal communication)

Liveweight (kg)	Ration M/D (MJ ME per kg DM)				
	9	10	11	12	13
	Crude protein (proportion of ration DM)				
100	-	-	0.18	0.21	0.24
200	-	0.13	0.14	0.15	0.16
>200	0.11	0.12	0.13	0.14	0.15

Note: Missing values indicate the M/D is too low to meet the ME requirement.

Rate of gain certainly affects the daily requirement for protein, but at a given ration M/D cattle growing faster have a greater feed intake which provides additional protein for the higher gain. More important still, faster growth is usually obtained by feeding higher M/D rations for which higher CP proportions are recommended. These higher M/D rations also have a higher proportion of FME in the ME which aids protein utilisation.

The highest CP proportions are required by young cattle in the 100kg liveweight category. It should be noted that the tabulated values for M/Ds of 12 and 13 are well above the present protein recommendation of 0.18 (18%). It is at this young age that DUP requirements are highest but there is also a substantial ERDP requirement, especially as the CP proportion increases at higher M/D values.

By the time cattle reach 200kg liveweight, CP proportions are little higher than at greater weights, which themselves can be combined for practical purposes. The CP proportions in Table 4.1 at these higher weights are close to existing recommendations. Again it is necessary to stress the importance of choosing ingredients which supply plentiful ERDP.

Cattle type affects the daily protein requirement and at 100kg liveweight there is a case for recommending a higher CP proportion for continental breed types and bulls than for British breed types, steers and heifers. This proposition has not been followed in Table 4.1 which presents CP proportions calculated for continental breed/cross bulls, on the grounds that steers and heifers and breed types with a lower rate of lean deposition are not likely to be fed high energy diets with M/Ds of 12 or 13 MJ/kg DM.

SUCKLER COWS

Crude protein proportions required by suckler cows are shown in Table 4.2. Rations based largely or solely on grass silage will often have an apparent surplus of CP.

Table 4.2   Recommended crude protein contents in diets for housed suckler cows

	Daily weight loss (kg)	Milk yield (kg)	Crude protein (proportion of DM)
Spring calver			
Pre-calving	-0.50	0	0.12
Suckling	-0.50	10	0.15
Autumn calver			
Pre-mating	-0.25	10	0.15
Post-mating	-0.50	7	0.15

Rationing Beef Cattle

Chapter 5

Vitamins and Minerals

Adequate levels of essential vitamins and minerals are just as important to cattle performance as the energy and protein of the diet. Indeed in cases of deficiency performance may be completely disrupted or the cattle may die. Allowances presented here are based on the recommendations of MAFF *et al.* (1984).

VITAMINS

Once rumen function is established cattle become largely independent of B vitamins, though rumen synthesis may be impaired unless the cobalt level of the diet is at least 11 mg per kg DM (parts per million - ppm).

Cattle fed green feeds and exposed to sunlight are usually well supplied with the major vitamins A and D. However, the requirements of housed cattle must be met and allowances for growing cattle and suckler cows are presented in Table 5.1.

The vitamin A allowances are lower than former estimates which were considered by MAFF *et al.* (1984) to be over-generous.

Further, to guard against the remote possibility of pregnant women ingesting toxic levels of vitamin A, a voluntary code has been agreed which limits inclusion rates to not more than 12000 international units (iu) per kg in concentrates for growing cattle and 10000 iu in concentrates for milking cows.

For home mixing of concentrates it is only necessary to include a vitamin supplement for calves or where housed cattle are fed high concentrate mixes, eg in cereal beef production.

Table 5.1   Recommended allowances for vitamins A and D
(Based on MAFF *et al.* 1984)

Liveweight (kg)	Vitamin A (iu per day)		Vitamin D
Growing cattle			
100	6500		600
200	13000		1200
300	20000		1800
400	26500		2400
500	33000		3000
Suckler cows			
	Pregnant	Lactating	
450	44500	100000	4500
550	54500	120000	5500
650	64500	145000	6500

MINERALS

Requirements for the major minerals: calcium (Ca), phosphorus (P) and magnesium (Mg) are presented in Table 5.2. Contents in feeds are presented in Table 3.1 (Chapter 3).

Meeting the Mg requirement of lactating suckler cows is especially important to minimise the risk of hypomagnesaemic tetany (grass staggers).

Copper (Cu) is the trace mineral most likely to cause problems and and on chalk, limestone and rain-leached soils deficiency should always be suspected in cases of ill thrift. Often the coats of black haired cattle become brindled. It may be more practicable to prevent the condition with Cu injections rather than relying on dietary supplementation. The recommended Cu level in cattle diets is 12 mg per kg DM.

Table 5.2 Recommended allowances for major minerals
(Based on MAFF *et al.* 1984).

Liveweight (kg)	Daily gain of growing cattle (kg)								
	0.6			1.0			1.4		
	Ca	P	Mg	Ca	P	Mg	Ca	P	Mg
				(g per day)					
Growing cattle									
100	20	8	3	31	12	5	43	17	6
200	22	11	5	32	15	6	43	19	7
300	25	15	7	36	19	8	47	23	11
400	30	25	9	40	28	10	50	32	11
500	32	31	10	41	34	12	51	38	13
Suckler cows									
	Pregnant			Early to mid lactation			Mid to late lactation		
450	32	29	11	43	38	15	35	33	13
550	36	36	13	47	45	17	39	40	15
650	39	43	14	50	52	19	42	47	17

On some clay soils a form of Cu deficiency called "teart" is induced by a high level of molybdenum, which causes elevated Cu excretion.

Selenium (Se) and vitamin E act together in an enzyme system which prevents muscular dystrophy and other wasting conditions. The margin between selenium deficiency (below 0.1 mg per kg DM) and toxicity (above 0.3 mg per kg DM) is very narrow.

For home mixing a proprietary powdered mineral mix can either be incorporated into the concentrate or, as a second best, top dressed onto the feed in the trough.

Chapter 6

Rationing

The procedure for ration formulation used here has the following steps:

- Set a target daily gain.
- Establish the DM, ME and CP contents of the feed ingredients.
- Predict the M/D of the final ration.
- Determine the daily ME requirement and the required CP level in the ration.
- Predict potential forage intake and calculate the amount of concentrate supplement required.
- Compile the daily ration.
- Draw up a feed budget for the whole feeding period.

Worksheets for growing cattle and suckler cows are presented in Appendix 3. Please photocopy these for your personal use.

GROWING CATTLE

The steps used in rationing growing cattle are presented in detail below, using as an example the winter finishing of 40 Charolais cross steers fed grass silage supplemented with concentrates.

The way in which the rationing procedure is presented here to explain the logic of each step, may make it seem cumbersome. However, its essential simplicity soon becomes evident when actually formulating rations using the worksheet. With practice, setting performance targets, formulating a ration and drawing up a feed budget can be completed in a few minutes.

(a) Cattle performance

Ideally weigh the cattle. Alternatively use the weight at purchase but beware of relying on visual estimates that are notoriously unreliable.

Use previous farm experience of the guidelines in Figure 1.1 or Table 1.1 to set targets for cattle performance, including daily gain and final weight. Calculate the average weight of cattle over the feeding period, total liveweight gain and the duration of the feeding period. Later the feeding period will be used in drawing up the feed budget - the total amount of each feed ingredient needed.

If the total weight gain is more than 100 kg or the feeding period lasts more than three months, it is advisable to work out rations for two or more sub periods. For the purposes of illustration, a single ration for the whole feeding period is presented here.

Daily gain	*1.0* kg
Start weight	*375* kg
Final weight	*525* kg
Average weight	*450* kg (start + final ÷ 2)
Total gain	*150* kg (final - start weight)
Feeding period	*150* days (total gain ÷ daily gain)

(b) Feeds

List the feeds with their DM, ME and CP values from a feed analysis or using data from Table 3.1. If necessary, convert the DM and CP to decimal proportions.

Ration ingredient	DM (Decimal)	ME (MJ/kg DM)	CP (Decimal)
Silage	*0.25*	*11.0*	*0.165*
Rolled barley	*0.86*	*13.3*	*0.129*

If two or more forages or concentrate ingredients are to be fed, it is necessary to decide on appropriate ME values and DM intake assumptions. Usually the concentrate ingredients are mixed and an average ME value can be calculated quite simply. With two forage sources, one may be fed as a fixed amount with the other fed to appetite or the two forages may be mixed in predetermined proportions. If the second forage is to be fixed in amount, it is accurate enough for practical purposes to deduct that amount from the potential DMI of the second forage (Table 2.1).

If the forages are mixed, work out the average ME content and decide on an appropriate potential DMI. In either case decide what is the likely concentrate substitution rate based on values in Table 2.1. If it all sounds rather approximate, that is the nature of rationing.

(c) Ration M/D

Predict the final ration M/D using the guidelines in Table 3.2.

Daily gain (kg)	Predicted M/D (MJ ME/kg DM)
1.0	11.0

(d) ME and CP requirements

Look up the daily ME requirement in Table 3.3 for the appropriate cattle type, average liveweight and M/D (if the M/D is between whole numbers estimate the ME requirement by inspection of neighbouring values). Look up the content of CP needed in the ration DM (Table 4.1).

Average Liveweight (kg)	Daily gain (kg)	ME requirement (MJ ME/day)	Ration CP content (Decimal proportion)
450	1.0	97	0.13

Now a daily gain target has been set and the ME and CP requirements to support it have been established. The next step is to formulate the daily ration, starting with a prediction of potential forage intake.

(e) Potential forage intake

Look up the potential forage intake in Table 2.1, making any adjustments for silage type or the body condition of the cattle. If more than one forage is fed, or if other bulk products are included in the ration, e.g. potatoes, make an estimate of the likely total potential forage intake. Also look up in Table 2.1 the concentrate substitution rate, ie the reduction in forage DM intake for each kg of concentrate DM fed.

Potential forage intake   ...*7.9*...kg DM/ day

Concentrate substitution   ....*0.5*...kg reduction in forage DM per kg concentrate DM

(f) ME from ad libitum forage

Work out how much ME would be supplied by silage alone at the potential intake.

Forage	ME content (MJ ME/kg DM)	x	Potential intake (kg DM/day)	ME supplied (MJ ME/day)
Silage	*11.0*		..*7.9*...	..*87*..

(g) Is there a shortfall of ME from forage?

Calculate if there would be a shortfall of ME from forage if it was fed alone at potential intake.

ME required (d)   ...*97*....MJ/day

ME from forage alone (f)   ...*87*....MJ/day

Shortfall   ....*10*...(d) - (f)

(h) Calculate the net value of 1 kg concentrate DM

Work out the net value of 1 kg concentrate DM, taking account of concentrate substitution. This is an extremely useful figure that will be used in step (i) to calculate the daily allowance of concentrates. This net value can be used without further calculation to explore concentrate requirements at different levels of gain.

1 kg concentrate DM supplies	...*13.3*...MJ ME
Less reduced forage DMI .*0.5*.kg	
@ ..*11*.MJ ME/kg DM =	..*5.5*....MJ ME
Net value of 1 kg concentrate DM	...*7.8*...MJ ME/kg

(i) Daily concentrate requirement

Work out the daily concentrate DM requirement simply by dividing the ME shortfall from forage alone (g) by the net value of 1 kg concentrate DM. Then the ME provided by the concentrate allowance is DM multiplied by its actual ME content from (b).

Shortfall(g) ..*10*..MJ ME ÷ Net value (h) *7.8*.MJME/kg DM

= ..*1.3*.kg concentrate DM x *13.3*. MJ ME/kg DM(b)

= ...*17*.MJ ME/day

(j) Actual forage intake

Calculate the ME that needs to be supplied by the forage and the resulting DMI.

ME requirement (d)	..*97*..MJ ME/day
Less concentrate ME (i)	...*17*.MJ ME
= Forage ME required	..*80*.MJ ME/day
÷ Forage ME/kg DM (b)	*7.3*.kg DM/day

(k) Check actual M/D

Check that the predicted M/D (c) is within 0.5 of the actual value. If not, go back to step (c) and use the value calculated here to look up a new ME requirement and recalculate steps (d), (g), (i) and (j).

Feed ingredient	DMI (kg DM/day)	ME (MJ/day)
Silage	7.3 (j)	80 (j)
Rolled barley	1.3 (i)	17 (i)
Totals	8.6	97

Actual M/D is total ME intake ÷

total DMI =   11.3 MJ ME/kg DM

Predicted M/D (c)   11.0 MJ ME/kg DM OK?

(l) Check CP

Check that the CP proportion in the DM conforms closely with the requirement in (d). If it is more than 0.01 too low, the usual solution is to increase the CP in the concentrate by juggling with ingredients. If it is too high, there may be scope for saving CP.

Feed ingredient	DMI (from k)	CP (DMI x Decimal CP in feed DM from b)
Silage	7.3	1.20
Rolled barley	1.3	0.17
Totals	8.6	1.37

Total CP ÷ Total DMI = 0.159 Decimal CP content OK?

(m) Daily ration

Compile the daily ration as fresh weights by **dividing** the daily DMI by the DM contents of the feed ingredients. For example: 6 kg silage DM ÷ 0.22 DM content = 27.3 kg fresh weight.

Feed ingredient	DMI (kg/day)	÷Decimal DM content (b)	=	Fresh weight (kg/day)
Silage	7.3	0.25		2.9
Rolled barley	1.3	0.86		1.5

At this stage the daily feed cost can be calculated to assess the cost per kg of gain.

(n) Feed budget

Work out the feed budget for the whole period by multiplying the daily ration for each feed ingredient by the number of days the cattle are fed to calculate total quantities required per head. Then multiply by the number of cattle to find total feed requirements.

Feed ingredient	Daily ration (kg)	xDays	=Feed/head (kg)	xCattle	= Total feed (t)
Silage	29	150	4350	40	174
Rolled barley	1.5		225		0.9

PREDICTING LIVEWEIGHT GAIN

Sometimes it is necessary to predict the liveweight gain supported by a stated daily ration. This is dome by working out the ME intake and calculating the M/D of the ration and then looking up in the appropriate section of Table 3.3 at the liveweight of the cattle and the M/D of the ration what the gain should be. For example:

Hereford x Holstein-Friesian steers weighing 400 kg liveweight
Daily ration: 25kg of 22% 11ME silage and 3kg of 86% 12.5ME concentrates
Silage   25kg x 0.22DM = 5.5kgDM @ 11MJ ME/kgDM   = 60.5MJ ME
Concentrates   3kg x 0.86DM = 2.58kgDM @ 12.5MJ ME/kgDM = 32.3MJ ME
Totals   8.08kg DM   92.9MJ ME

M/D = 92.8/8.08 = 11.5 MJ ME/kg DM

Look up Table 3.3 Section C at 400 kg and estimate values between M/D 11 and 12. The ration will support a daily gain of approximately 1.0 kg.

SUCKLER COWS

Formulating rations for suckler cows using the worksheet in Appendix 3 follows essentially the same procedure as for growing cattle. In fact it is simpler because the ME allowances in Table 3.4 incorporate performance targets. Moreover, the M/D prediction for the ration is not usually critical on the relatively low quality rations fed. Often dry cows are fed restricted amounts of forage with an M/D above the minimum guideline of 8.5 MJ ME/kg DM. Usually straw is fed to appetite.

Chapter 7

Rationing Tactics

FEED BUDGET

The last but by no means the least important step in formulating a ration is to calculate a feed budget - the total quantity of each feed ingredient needed to sustain the whole feeding period. All too often there is insufficient forage in store to carry out the original feeding plan. Late in winter is not the time to find this out. Either, it precipitates the enforced sale of cattle as stores when the market may be depressed. Or, an expensive rescue package has to be mounted with expensive purchased feeds. If a potential problem is identified at the start of the winter, contingency plans can be drawn up to deal with it.

RATIONING TACTICS FOR FINISHING CATTLE

There is considerable scope for manipulating cattle performance to alter the amount of forage used or to manipulate slaughter weight and date in response to market requirements.

The principle that makes this possible is that raising daily gain by increasing the daily allowance of concentrates, not only reduces the daily forage intake, but also shortens the feeding period because cattle finish more rapidly to a lower slaughter weight. There is a major saving of forage with, at most, only a small increase in total concentrate consumption.

The example in Table 7.1 shows just how much room for manoeuvre exists. Charolais cross steers weighing 385 kg at the start of the winter are finished on a well-made 25% DM silage with 11 MJ ME/kg DM supplemented with rolled barley containing 13 MJ ME/kg DM. Raising the gain from 1.0 kg per day to 1.2 kg per day shortens the feeding period to such an extent that there is a major saving in silage of more than 3 tonnes of silage per head at the cost of only an extra 50 kg rolled barley.

Table 7.1   Rationing tactics for Charolais cross steers.

Daily Gain (kg)	**1.0**	**1.1**	**1.2**
Start weight (kg)	385	385	385
Slaughter weight (kg)	550	515	480
(From Figure 1.1)			
Total weight gain (kg)	165	130	95
(Slaughter – Start)			
Finishing period (days)	165	118	79
(Total gain ÷ Daily gain			
Daily ration (kg)			
Silage	30	27	24
Rolled barley	1.5	2.7	3.8
Feed budget (tonnes)			
Silage	5.0	3.2	1.9
Rolled barley	0.25	0.32	0.30

RATIONING TACTICS FOR SUCKLER COWS AND STORES

With suckler cows and stores there is not the option to alter the duration of the feeding period. Indeed, rough weather can extend the winter by several weeks. So, if forage is short a way has to be found of eking out supplies. Invariable there is an increase in feed costs but that is preferable to allowing cows to lose excessive weight and body condition which may compromise the next calf crop.

With suckler cows the point has already been made that it is more efficient to creep feed calves directly, rather than giving cows extra feed to try and increase milk yield with the inefficient two-stage process of converting feed to milk then milk to calf gain.

If forage is short, it may be possible to purchase silage, particularly as big bales, but is the purchased silage competitive in price with an alternative ration of silage, straw and concentrates? The straw may be better value for money, even if it has to be upgraded by treatment with ammonia. The higher DMI of treated straw saves concentrates which helps to defray the cost of treatment.

Table 7.2 presents a range of alternative rations for 500 kg suckler cows losing 0.25 kg liveweight and producing 10 kg milk daily. Over a three-month feeding period, at the expense of 250 kg concentrates and 225 kg straw, the silage requirement can be almost halved compared to feeding an all-silage diet, albeit probably at greater cost.

Table 7.2   Alternative rations for suckler cows.

Liveweight: 500kg
Liveweight loss: 0.25 kg/day
Milk yield: 10 kg/day

Feeds (kg)	**Daily ration** (kg)	**90-day feed budget** (kg)
Silage	45	4050
Silage	25	2250
Straw	2.5	225
Concentrates	3.0	270
Silage	15	1350
Straw	3.5	315
Concentrates	4.5	405
Silage	5.0	450
Straw	4.5	378
Concentrates	6.0	540

It bears repeating that the time to study feed budgets to compare ration options is before cows are housed when forage looks as though it may be in short supply, not late in the winter when the situation is already desperate.

The same arguments apply to alternative winter rations for store cattle. The is also the possibility of reducing winter gain slightly, knowing that compensatory growth during the next grazing season will claw back the difference, though probably at the cost of a later start to marketing off grass.

Chapter 8

Management of Cattle and Feeds

Time has been well spent setting performance targets for cattle, formulating rations and working out feed budgets. These are, however, just the preliminaries of profitable beef production.

Day-by-day throughout the feeding period, it is essential to present the cattle with the right amount of fresh feed and to have them penned so that they are able to consume it. A simple means of monitoring progress is also required to check progress and to provide information so that rations can be revised when necessary.

CATTLE MANAGEMENT

When cattle are introduced to different rations it takes the rumen microrganisms about three weeks to adapt. Therefore, it is important to make diet changes as gradually as possible. Young calves turned out to graze for the first time benefit from a short period of concentrate supplementation until they become accustomed to grazing. Similarly, grazed cattle to be finished during the winter benefit from being introduced to their winter ration gradually before housing whilst still grazing. They need supplementary feed anyway from mid-August to counter the fall in grass availability and quality.

If there is a large increase in the amount of concentrates fed, or if cattle are being switched to an all-concentrate diet, gradual change is essential. Otherwise, dominant cattle in the group will gorge themselves on the concentrates with a dire risk of fatal bloat and acidosis. Even if affected cattle recover, their performance is likely to be permanently sub-normal.

Bloat and acidosis are a continuing threat to cattle fed an all-concentrate diet. It helps to have bright straw available so that the cattle can consume a little roughage. Anything that interrupts feeding - a soiled feed trough, feed bridging in a hopper, feed running out or the water supply failing for several hours in winter - puts the cattle at risk when feeding starts again.

Housed cattle should be inspected at least twice daily watching for any signs of disease, especially pneumonia. Suspect cattle may merely hang back from the trough at feeding time or they may be at a more advanced stage of symptoms with runny noses or coughing. Early treatment under veterinary guidance is essential to avoid heavy infection that, even if it is cured, has lasting detrimental effects on performance.

FEED MANAGEMENT

Feeds need to be managed with the same care as the cattle. Poorly made silage and stale or musty feeds will never deliver peak performance, whatever their compositional analysis. It is essential to keep fresh feed in front of the cattle and to remove stale or soiled feed from troughs or feed hoppers each day.

Low DM silage carted straight from the field without wilting is eaten in smaller quantities than drier silage and the copious effluent not only wastes feed value but is a dangerous pollutant. In theory the effluent can be fed but few farmers are equipped to do so. Particular care is needed during cutting and carting grass to avoid soil contamination which reduces intake.

Quite apart from feeding value, pollution regulations are a clear incentive to cut in dry weather and employ a 24-hour wilt, aiming for a silage with 25% DM that produces little or no effluent. For wetter silages there may be a case for adding an absorbent such as dried sugar beet pulp to minimise effluent loss. At 20% DM the beet pulp is mixed at 50 kg per tonne of grass.

During making, the silo should be filled rapidly, consolidated and sealed daily with a plastic sheet to prevent spoilage. In poor weather an acid or effective inoculant additive is an aid to good fermentation and suppresses undesirable butyric acid fermentation that spoils silage. Many farmers use an additive routinely as an insurance policy.

Some well made silages have very low pH values (3.5 to 4.0) which cattle may find unattractive and which can cause acidosis. It may be necessary to supplement acid silage with a buffer such as sodium bicarbonate.

A problem with grass and maize silage is spoilage at the face due to aerobic secondary fermentation. It is worst in warm weather. It helps if silage clamps are narrow and a block cutter is used to remove silage so that the face stays compacted.

It is notoriously difficult to make good quality hay over 80% DM unless there is a prolonged hot, dry spell of weather. Heavy first cuts are a particular problem. With damp hay moulding is a problem which reduces intake and may cause health problems in cattle and stockworkers alike. Many farmers who used to make hay have switched to big bale silage.

Straw for feeding should be baled and carted while it is dry and bright in colour so that it does not become musty. If the straw is to be treated in a stack with ammonia, it is best carted soon after harvest when the weather is still warm and the reaction works well. Feed grain stored at 85% DM is likely to shatter when rolled, producing an undesirable dusty feed. Rolling works best with moist grain stored in a tower or treated with propionic acid.

It is sensible with purchased concentrates to take advantage of cheaper bulk delivery. The feed might be stored in a grain trailer not needed for transport over the winter. If the feed has to be stored on the floor, it is wise to take the smallest discounted delivery. The concentrate needs to be kept dry and spoilage by birds and rodents prevented.

Home mixers in the UK producing medicated feeds, for example containing growth promoters, are obliged to register under what is known as 'Category B' which lays down procedures for feed mixing, labelling and keeping records.

FORAGE SUPPLIES

Interpreting the feed budget depends on knowing how much silage or other forage is in store. Stocktaking at the start of the winter can prevent unpleasant and expensive surprises later on. Assessing quantities is relatively simple for hay, straw and big bale silage where bales of known average weight can be counted.

Assessing the amount of silage in a clamp is more difficult but should be done systematically and not, as happens all too often, be the subject of a wildly optimistic guess. Measure the height of the silage face, the width of the clamp, its length at full height and the length of the sloping wedge. Volume is then:

(Height x width x length at full height) + ($^{1}/_{2}$ length of sloping wedge x height x width)

To estimate the weight of silage multiply volume by silage density in the clamp, which for short-chopped material in a clamp with a face 3 metres high, is about 0.9 tonnes per cubic metre at 20% DM, 0.75 tonnes at 25% DM and 0.60 tonnes at 30%.

Whole crop cereal treated with urea has a high DM content of 50 to 60% but a low density of 0.35 to 0.40 tonnes per cubic metre. Nevertheless the weight of DM per cubic metre is similar to grass silage.

BUILDINGS

Cattle cannot perform to potential unless they are penned cleanly and comfortably with easy access to feed so that even the more timid animals are able to consume the ration set for them. In general performance is better for cattle housed in small groups. The UK safety code for bull beef, for example, proposes an upper limit of 20 bulls in a group. Inevitably, the smaller the group the greater is the cost of pen divisions and providing water.

Table 8.1 shows space allowances, excluding troughs, taken from the Farm Assured British Beef and Lamb (FABBL) farm assurance specification. There are reservations about the welfare of cattle housed on slatted floors. For example, the UK welfare code recommends straw bedding for cows and that they should not be kept on totally slatted floors. Nevertheless, where buildings have slatted floors, the cattle should be stocked heavily enough so that the slats self-clean.

Cubicles that are so widely used by dairy farmers are suitable for dry, pregnant suckler cows but with autumn calving suckler cows there is the problem of what to do about the calves. Cubicles are not very suitable for growing cattle because it is difficult to adjust the dimensions as cattle grow larger. Also male cattle urinate in the middle of the cubicle bed and make it damp and riding bulls can damage the cubicle structure. Cubicle dimensions recommended by FABBL for growing cattle are presented in Table 8.2. Heavy suckler cows over 600 kg need a cubicle width of 2.2 metres and length of 2.3 metres.

The feed space required at the trough or feed barrier is dictated by feed type. Most commonly cattle are fed rationed concentrates once or twice daily with forage to appetite. In this case all the cattle need to be able to feed at once. Recommended space allowances are shown in Table 8.3. Less space is needed if the feed ingredients are mixed as a total mixed ration and fed ad libitum. Then it is satisfactory if two thirds of the cattle are able to feed at once and the feed is distributed at least twice daily. In top US feedlots feed is distributed three times daily to ensure that it is always fresh.

Table 8.1 Space allowances for housed cattle excluding troughs. Source: FABBL

Liveweight (kg)	Square metres per head excluding troughs	
	Straw bedding	Slatted floors
200	3.0	1.1
300	3.4	1.5
400	3.8	1.8
500	4.2	2.1
600	4.6	2.3
700	5.0	2.5

Table 8.2 Cubicle dimensions for beef cattle. Source: FABBL

Liveweight (kg)	Cubicle length (m)	Clear width (m)*
75 to 150	1.2	0.60
150 to 250	1.5	0.75
250 to 375	1.7	0.90
Over 375	2.1	1.10

*Width between partitions

Table 8.3 Trough space required for all cattle to feed together. Source: FABBL, Author

Liveweight (kg)	Trough space (mm)
Growing cattle	
<250	400
250 to 450	500
>450	600
Suckler cows	
500 to 600	675
>600	700

RATION MONITORING

Making sure that the ration is fed by stockworkers, that cattle consume it and that performance stays on target is a chore but essential to profitable production.

Commonly performance is below target because of exaggerated ideas about how much silage is in a trailer load or home-mixed concentrates in a bag. Check weighing sample loads to establish quantities is well worth the effort. Farmers with feeder wagons fitted with load cells are in the best position to monitor quantities fed. But cost is beyond the means of most beef producers, except those with very large enterprises.

Having offered the right amount of feed, do the cattle eat it? Reasons for lower than expected feed intake include errors in the prediction of potential forage intake, feed pushed away from a feed barrier may be out of reach of the cattle, there may be competition at the feed barrier or trough, or there may be health problems.

Once newly housed cattle have been given a few days to settle to their surroundings, it is worth trying to establish that feed intake is as planned. By the time that a weighing reveals that gains are below target, so much time may have been lost that costly changes are necessary to the ration to get the cattle back on course.

Nevertheless, regular weighing of growing cattle is important to establish the rate of gain being achieved. Action should be taken if gains are more than 0.1 kg per day above or below target or if gains are too variable. In any case a review of rations should be undertaken in mid winter.

Bibliography

ADAS (1984) *Energy Allowances and Feeding Systems for Ruminants.* Reference Book 433. HMSO, London.

AFRC (1990) *Nutrient requirements of ruminant animals: Energy.* Report No. 5. Nutrition Abstracts and Reviews (Series B), 60, 729.

AFRC (1992) *Nutrient requirements of ruminant animals: Protein.* AFRC Technical Committee on Responses to Nutrients. Nutrition Abstracts and Reviews (Series B), 62, 787.

AFRC (1993) *Energy and Protein Requirements of Ruminants.* CABI, Wallingford.

FABBL (1999) *Beef and Lamb Scheme Producers Manual.* FABBL/ABM, P.O. Box 165, Milton Keynes MK6 1PB.

MAFF, DAFS, DANI, UKASTA (1984). *Working party report. Mineral, trace element and vitamin allowances for ruminant livestock.* In: Recent Advances in Animal Nutrition Eds W. Haresign and D J A Cole. Butterworths, London.

Appendix 1
ME System for Growing Cattle

AFRC (1990) Model of ME requirements for growing cattle and assumptions adopted in this manual

$$\text{ME (MJ PER DAY)} = \frac{Em}{p} \; \ln \; \frac{B}{B - R - 1}$$

1. $E_m = C1 * 0.53 \, (W/1.08)^{0.67} + 0.0071 \, W$

   Where:

   E_m = maintenance energy (MJ)

   W = liveweight (kg)

   $C1$ = correction factor

   ARC(1990): $C1 = 1.0$ for steers and heifers

   1.15 for more active bulls

   This manual: $C1$ of 1.15 for household bulls not adopted.

2. $p = k_m * \ln (k_m/k_f)$

   Where:

   k_m (maintenance efficiency) = $0.35 q_m + 0.503$

   k_f (growth efficiency) = $0.78 \, q_m + 0.006$

   and:

   $q_m = \text{ME} / \text{GE}$

   GE: gross energy of the feed taken as 18.8MJ for mixed rations usually containing silage.

3. $\ln$ is the natural logarithm.

4. $B = \dfrac{k_m}{k_m - k_f}$

5. $R = \dfrac{EVg * \blacktriangle W}{Em} * C4$

Where:
EVg is the energy value of grain

$$= \frac{C2\,(4.1 + 0.0032W - 0.000009W^2)}{1 - C3(0.1475\blacktriangle W)}$$

C2 is a correction factor for breed maturity type and sex
AFRC (1990) classification of breed types:

	Bull	Steer	Heifer
Maturity - early	1.00	1.15	1.30
medium	0.85	1.00	1.15
late	0.70	0.85	1.00

An alternative classification is preferred in this manual:

	Bull	Steer	Heifer
Continental Cross	0.8	0.9	1.0
British Cross	0.9	1.0	1.1
Fresian/Holstein	1.0	1.1	1.2

$$C3 = 1 \text{ when } \frac{ME\ intake \ast k_m}{Em} > 1$$

This is the case at the rates of gain of 0.5kg per day or more listed in this manual.

$\blacktriangle W$ is the daily gain (kg)

C4 is a correction factor to allow for overall biases in the calculation of R.

AFRC (1990) : C4 bulls/steers 1.15, heifers 1.10

This manual: lower C4 for heifers not adopted;
                 bulls, steers and heifers all 1.15

Note that in ARC (1990) there is confusion over how the C4 correction is used. In the definition of formulae it is applied to the calculation of R, but in the worked example it is applied to the final calculation of the ME requirement. A personal communication from the working party confirms that C4 should be applied to the calculation of R.

6.   A safety margin of 1.05 has been applied to the calculation of daily ME requirements.

Appendix 2
ME System for Suckler Cows

**AFRC (1990) model of ME requirements for cows and
assumptions adopted in this manual**

Lactating cows, non-pregnant or early pregnancy

$$ME(MJ/DAY) = CL\left[\frac{E_m}{k_m} + Y*\frac{EV_1}{k_1} + \blacktriangle W*\frac{EV_g}{k_1}\right] C5$$

Pregnant, dry cows

$$ME(MJ/DAY) = CL\left[\frac{E_m}{k_m} + \frac{E_c}{k_c} + \blacktriangle W*\frac{EV_g}{k_g}\right] C5$$

1.   CL is a level of feeding correction which is not
     considered necessary for relatively low performance
     suckler cows losing weight on winter rations.

     CL for lactating cows:

$$CL = 1 + 0.018\left[\frac{Y*EV_1}{k_1} + \frac{W*EV_g}{k_g}\right]*\frac{k_m}{E_m}$$

     CL for pregnant cows:

$$CL = 1 + 0.018\left[\frac{E_c}{k_c} + \frac{E_g}{k_g}\right]*\frac{k_m}{E_m}$$

2. $Em = 0.53\left[\dfrac{W}{1.08}\right]^{0.67}$ $+0.0091\ W$ (lactating cows)
     $+0.0071\ W$ (pregnant cows)

3.   Y is milk yield (kg/day).

4. $EV_1(MJ/kg)$ is the energy value of milk with
     constituents measured as g/kg.

     = 0.0384 Fat + 0.0223 Protein + 0.01992 Lactose - 0.108

5. ΔW is daily liveweight gain or loss (kg)

6. EV_g is the energy value of liveweight loss, taken as 27.36 MJ/kg in lactating and 26 MJ/kg in pregnant cows.

7. E_c is the energy value of the conceptus and is taken here as a constant 3 MJ/kg over the last five months of pregnancy. With a k_c of 0.133, the daily requirement to support pregnancy is 22.6 MJ ME.

8. Efficiencies:

    $km = 0.35q_m + 0.503$ (where $q_m = ME/GE = ME/18.8$)

    $k_1 = 0.35\ q_m + 0.42$

    $k_g = k_1$ for liveweight gain in lactating cows

    $k_g = k_1/0.8$ for liveweight loss in lactating cows

    $k_g = k_f = 0.78q_m + 0.006$ for grain in pregnant cows

    $k_g = 0.665$ for liveweight loss in pregnant cows

    $k_c = 0.133$

9. C5 is a safety margin, taken as 1.05 for lactating cows. No safety factor has been adopted for dry, pregnant cows.

Appendix 3a

Ration Worksheet - Growing Cattle

BASIC INFORMATION

1 No of cattle		6 Predicted M/D (MJ/kg DM		*Table 3.2*
2 Breed type/sex		7 ME requirement (MJ/day)		*Table 3.3*
3 Daily gain (kg)		8 Required CP in diet (Decimal)		*Table 4.1*
4 Start weight (kg)		9 Average weight (kg)		*Start + Final ÷ 2*
5 Final weight (kg)		10 Total liveweight gain (kg)		*Final - Start*
		11 Feeding period (days)		*Total gain ÷ Daily gain*

FEED INGREDIENTS

12	13	14	15
Feed ingredients	DM (Decimal)	ME (MJ/kg DM)	CP (Decimal)

Look up Table 3.1 or a feed analysis for values in columns 13, 14 and 15

FORAGE POTENTIAL

16 Potential forage DMI (kg)	17 Potential forage ME (MJ/day)	18 Forage ME shortfall (MJ/day)?	19 Concentrate substitution (kg forage/kg concentrates)

Column 16:   Look up value in Table 2.1
Column 17:   Potential forage DMI from 16 x forage ME from 14
Column 18:   Total ME requirement from 7 minus potential forage ME from 17
Column 19:   Look up reduction in forage DMI per kg concentrates fed in Table 2.1

RATION FORMULATION

20	21	22	23	24
Feed ingredients	Net value of 1 kg concentrates (MJ ME/kg DM)	Concentrate DMI (kg/day)	Concentrate ME (MJ ME/kg DM)	Actual forage DMI (kg/day)

Column 21: The ME of 1 kg concentrates from 14 minus substitution from 19 x forage ME from 14

Column 22: ME shortfall from 18 ÷ net value of concentrates from 21

Column 23: Concentrate DMI x ME of concentrates from 14

Column 24: Total ME from 7 minus concentrate ME from 23 ÷ forage ME from 14

CHECK RATION M/D AND CP CONTENT

25	26	27	28
Feed ingredients	DMI (kg/day)	ME (MJ/day)	CP (ration proportion)
Forage			
Concentrates			
Totals			
M/D (MJ ME/kg DM)	-		-
CP (ration proportion)	-	-	
		M/D OK?	CP content OK?

Column 26: Concentrate DMI + forage DMI

Column 27: Concentrate ME value from 23 + forage DMI from 24 x ME value from 14. Divide the total ME by the total DMI from 26, calculate the actual M/D and check that it is within 0.5 of the predicted value in 6. If the difference is more go back to 6 and enter the M/D from 27, enter a revised ME requirement in 7 and revise values in columns 18, 22, 23, 24, 26 and 27

Column 28: Concentrate DMI from 22 x concentrate CP value from 15 + Forage DMI x forage CP value from 15. Divide by the total DM and check that the actual ration CP content is not more than 0.01 below the requirement in 8, or above. If above, are protein savings possible?

DAILY RATION AND FEED BUDGETS

29	30	31	32	33
Feed ingredients	Daily Concentrates	Daily Forage	Feed budget per head	Total feed budget
	(kg fresh weight)	(kg fresh weight)	(tonnes)	(tonnes)

Column 30: Concentrate DMI from 22 ÷ Concentrate DM content from 13

Column 31: Forage DMI from 24 ÷ Forage DM content from 13

Column 32: Daily concentrates from 30 x feeding period from 11. Daily forage from 31 x feeding period. If required calculate daily feed cost and feed cost per kg gain.

Column 33: Feed budget per head x number of cattle from 1

Appendix 3b
Ration Worksheet - Suckler Cows

BASIC INFORMATION

1 No of cows		6 Predicted M/D (MJ/kg DM	*Table 3.2*
2 Average weight (kg)*		7 ME requirement (MJ/day)	*Table 3.4*
3 Autumn condition score		8 Required CP in diet (decimal)	*Table 4.2*
4 Daily weight loss (kg)**		9 Feeding period (days)	
5 Daily milk yield (kg)**			

*   Actual or Table 1.4

** Table 3.4

FEED INGREDIENTS

10	11	12	13
Feed ingredients	DM (Decimal)	ME (MJ/kg DM)	CP (Decimal)

Look up Table 3.1 or a feed analysis for values in columns 11, 12 and 13

FORAGE POTENTIAL

14 Potential forage DMI (kg)	15 Potential forage ME (MJ/day)	16 Forage ME shortfall (MJ/day)?	17 Concentrate substitution (kg forage/kg concentrates)

Column 14: Look up value in Table 2.2
Column 15: Potential forage DMI from 14 x forage ME from 12

Column 16: Total ME requirement from 7 minus potential forage ME from 15. If there is an excess of ME formulate a ration including a cheaper lower quality forage e.g. straw

Column 17: Look up reduction in forage DMI per kg concentrates fed in Table 2.2

RATION FORMULATION

18	19	20	21	22
Feed ingredients	Net value of 1 kg concentrates (MJ ME/kg DM)	Concentrate DMI (kg/day)	Concentrate ME (MJ ME/kg DM)	Actual forage: DMI (kg/day)

Column 19: ME of 1 kg concentrates from 12 minus substitution from 17 x forage ME from 12

Column 20: ME shortfall from 16 ÷ net value of concentrates from 19

Column 21: Concentrate DMI from 20 x ME of concentrates from 12

Column 22: Total ME from 7 minus concentrate ME from 21 ÷ forage ME from 12

CHECK RATION M/D AND CP CONTENT

23	24	25	26
Feed ingredients	DMI (kg/day)	ME (MJ/day)	CP (ration proportion)
Forage			
Concentrates			
Totals			
M/D (MJ ME/kg DM)	-		-
CP (ration proportion)	-	-	
		M/D OK?	CP content OK?

Column 24: Concentrate DMI from 20 + forage DMI from 22

Column 25: Concentrate ME from 21 + forage DMI from 22 x ME value from 12. Divide the total ME by the total DMI from 24, calculate the actual M/D and check that it is within 0.5 of the predicted value in 6. If the difference is more go back to 6 and enter the M/D from 27, enter a revised ME requirement in 7 and revise values in columns 16, 20, 21, 22, 24 and 25.

Column 26: Concentrate DMI from 20 x concentrate CP value from 13 + Forage DMI from 22 x forage CP value from 13. Divide by the total DM and check that the actual ration CP content is not more than 0.01 below the requirement in 8. If above, are protein savings possible?

DAILY RATION AND FEED BUDGETS

27	28	29	30	31
Feed ingredients	Daily Concentrates	Daily Forage	Feed budget per head	Total feed budget
	(kg fresh weight)	(kg fresh weight)	(tonnes)	(tonnes)

Column 28: Concentrate DMI from 20 ÷ Concentrate DM content from 11

Column 29: Forage DMI from 22 ÷ Forage DM content from 11

Column 30: Daily concentrates from 28 x feeding period from 9. Daily forage from 29 x feeding period. If required calculate the daily feed cost.

Column 31: Feed budgets per head x number of cows from 1.